Contents

What are oceans and seas?

Oceans and seas are mighty bodies of salt water that cover nearly three-quarters of the Earth's surface. The edges of oceans are partly marked by the coasts of huge land masses called **continents**. They are also marked by underwater ridges of rock and by surges of flowing water called **currents**.

Seas are smaller than oceans and are mostly surrounded by land. They are often connected to oceans by narrow strips of water called **straits**. Some seas run between two strips of land and have large openings at each end. Others lie in scooped-out coastlines. We shall see where in the world these oceans and seas lie.

This is the Royal Princess. It is one of the biggest and most luxurious ocean liners in the world. It takes holiday-makers on cruises to many parts of the world. The oceans and seas have been used for travel, trade and exploration for thousands of years.

Oceans
and Seas

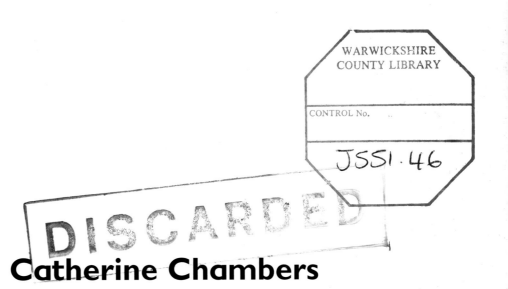

Catherine Chambers

Heinemann

First published in Great Britain by Heinemann Library,
Halley Court, Jordan Hill, Oxford OX2 8EJ,
a division of Reed Educational and Professional Publishing Ltd.
Heinemann is a registered trademark of Reed Educational & Professional Publishing Limited.

OXFORD MELBOURNE AUCKLAND
JOHANNESBURG BLANTYRE GABORONE
IBADAN PORTSMOUTH NH (USA) CHICAGO

Designed by David Oakley
Illustrations by Tokay Interactive Ltd and AMR
Originated by Dot Gradations
Printed in Hong Kong/China

05 04 03 02 01
10 9 8 7 6 5 4 3 2 1

ISBN 0 431 09850 6
This title is also available in a hardback library edition (ISBN 0 431 09843 3)

British Library Cataloguing in Publication Data

Chambers, Catherine
        Oceans and seas. – (Mapping earthforms)
        1. Marine ecology – Juvenile literature 2. Oceans – Maps –
        Juvenile literature
        I. Title
        577.7

Acknowledgements
The Publishers would like to thank the following for permission to reproduce photographs: Aspect Picture
Library: D Bayes p5; Bruce Coleman Limited: J Burton p18; Ecoscene: C Cooper p20, P Fernby p27; Oxford
Scientific Films: T Bomford p19, C Bromhall p7, P Parks p16, K Westerskov p24, S Winer p29; P & O: p4; G R
Roberts: p9; Still Pictures: B and C Alexander p23, K Andrews p10, D Hinrichson p12, D Watts p13, F Dott p21,
H Schwarzbach p25, P Glendell p26; Topham Picture Point: K Kasahara p14; UK Hydrographic Office (Crown
Copyright):

Cover photograph reproduced with permission of James L Amos and Still Pictures.

Every effort has been made to contact copyright holders of any material reproduced in this book. Any
omissions will be rectified in subsequent printings if notice is given to the Publisher.

For more information about Heinemann Library books, or to order, please phone ++44 (0)1865 888066, or send
a fax to ++44 (0)1865 314091. You can visit our website at www.heinemann.co.uk.

Any words appearing in the text in bold, **like this,** are explained in the Glossary.

How have oceans and seas formed?

Oceans and seas formed millions of years ago. Great splits in the Earth's crust separated the continents and left huge **basins**. These were filled with water that ran down from the mountains and hillsides in rivers and streams. As the rivers ran, they picked up salts and other chemicals from the rocks over which they flowed. The salts and chemicals were carried all the way to the sea, as they still are today. We will see how the water in the oceans is changing, and how the ocean floors are moving all the time.

What do they look like?

Oceans and seas have many different depths and colours. Their waves can lap the shores gently or they can rise to huge peaks that batter the coastline. We shall see what makes the oceans, seas and coastlines change their appearance.

Supporting life

Life on Earth began in the waters of our oceans and seas. Now, they are home to a great range of plant and animal **species**. Humans have lived on the shores for many thousands of years, eating the fish, **mammals** and plants that live in the waters. But we have also **polluted** the waters, threatening ocean life. We shall find out what the future holds for the oceans and seas of the world.

The oceans and seas have given us a rich supply of food. Shellfish are plentiful among the rocks and rock pools of Brittany in northern France.

Oceans and seas of the world

Counting the oceans

There are three major oceans of the world. These are the Atlantic, Pacific and Indian Oceans. Towards the **South Pole**, these oceans merge together in a **current** of water called the West Wind Drift. This is near Antarctica and many scientists call these waters the Antarctic, or Southern, Ocean.

Near the **North Pole**, the Atlantic reaches a circular stretch of water which many scientists call the Arctic Ocean. But some believe it is just part of the Atlantic Ocean. You will notice on the map that the Atlantic and Pacific Oceans are divided up into areas, such as the South Pacific.

Oceans cover different areas of **climate,** as you can see from the map. Many seas are too small to stretch through different climates. The Mediterranean Sea area all has roughly the same climate, which is often called 'Mediterranean'. But, as we shall see on pages 12–13, oceans and seas themselves help to form and change our climates more than any other force on Earth.

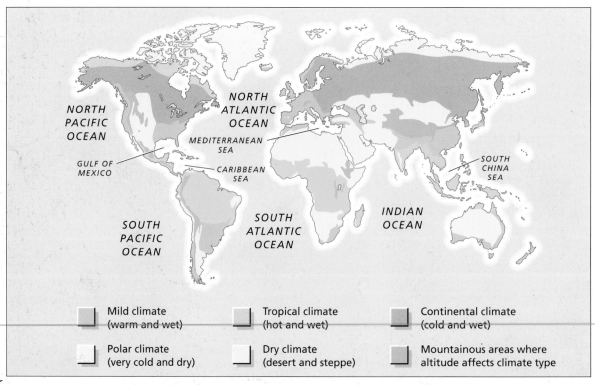

NORTH PACIFIC OCEAN

NORTH ATLANTIC OCEAN

MEDITERRANEAN SEA

GULF OF MEXICO

CARIBBEAN SEA

SOUTH CHINA SEA

SOUTH PACIFIC OCEAN

SOUTH ATLANTIC OCEAN

INDIAN OCEAN

Mild climate (warm and wet)

Tropical climate (hot and wet)

Continental climate (cold and wet)

Polar climate (very cold and dry)

Dry climate (desert and steppe)

Mountainous areas where altitude affects climate type

Sorting out the seas

Seas take many different shapes and forms. There sometimes seems to be no difference between a sea and a gulf or bay, which are scooped inlets near the shore, or between a sea and a channel, which is a strip of water between two coastlines. The Tasman Sea runs between the east coast of Australia and the islands of New Zealand. The ends of the sea are quite open, so it is like a channel. But the Black Sea is surrounded by land except for a very narrow strip of water called the Bosporus. In between these two examples is the Caribbean Sea, which is surrounded by the shores of Central America and a delicate string of islands. It is very similar to the nearby Gulf of Mexico.

◈ Sometimes, one sea leads into another. This is the Sea of Marmara. It lies between the Black Sea to the east and the Mediterranean Sea to the west.

How oceans and seas formed

The Earth was once a ball of hot gases. It is thought that about 4 billion years ago, the ball of gases cooled and made a solid mass of land called Pangaea. Volcanoes continued to throw out more gases and **water vapour**, forming clouds which shed rain. The Earth's **gravity** pulled the rain down mountainsides in streams and rivers. The rivers gathered into a huge **basin**, which became a massive acid ocean. Later, the waters became less acid and more salty – rather like the sea as we know it today.

Sixty million years ago, the great **plates** of rock on which Pangaea rested were pushed apart by the movement of hot **magma** under the Earth's crust. The land split up and different oceans and **currents** began to flow around the world.

You can see that the Earth once had a huge land mass called Pangaea. Then, a warm sea called Tethys separated the northern part, Laurasia, from the southern part, Gondwanaland. Different land masses continued to separate out or join together. Finally, Australasia and Antarctica split from each other, which allowed a cold current to circulate around the world.

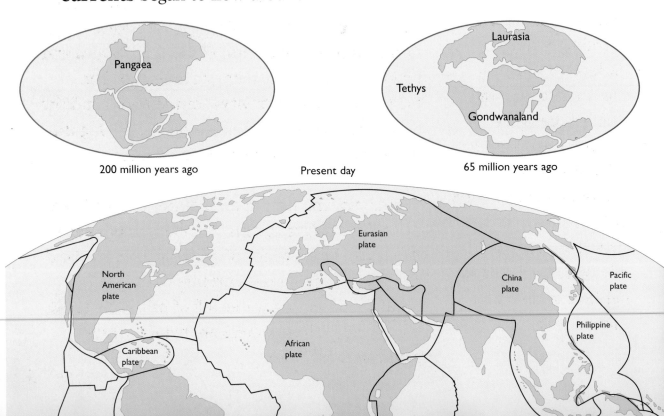

Pangaea

200 million years ago

Present day

Laurasia

Tethys

Gondwanaland

65 million years ago

Eurasian plate

North American plate

China plate

Pacific plate

Philippine plate

Caribbean plate

African plate

 When waves lap the shores they gradually wear away the cliffs on the coast. This leaves a long, flat **wave-cut platform** which leads to the cliff. The platform is full of rock, pebbles and sand. Some of this **debris** is knocked against the cliff, **eroding** it even more. This is known as **abrasion**. The waves cut bays into soft rock, leaving the **headlands** sticking out. These then get worn too. Sometimes a column of rock called a stack is left on its own. These stacks are called the Twelve Apostles and can be found on the coast of Victoria, Australia.

Moving waters

Very salty or cold waters sink, moving the water around them and below them. These massive movements of water are called underwater currents. The rotation of the Earth and winds blowing mainly in one direction above the water also make currents. These winds are called prevailing winds. They cause constant currents like the North Atlantic Drift and the Kuro Siwo in the North Pacific. Both of these are warm and make the lands near them warmer too. The magnetic pull – or gravity – of the Moon and the Sun on the water creates the tides, slopping the water from one side of a sea to another.

Winds blowing from the oceans to the land can bring destructive storms. But in parts of South America, strong winds often blow from the land to the sea. This causes currents deep in the ocean, which bring water full of **minerals** to the surface. These waters are helpful to sea plants and fish.

9

Looking below the surface

The ocean basin

The ocean **basin** goes down in steps. These get deeper as they get further from the **continents**. The first step is called the **continental shelf**. It runs from the shores of continents and into the oceans for an average of 75 kilometres (43 miles). In some places, though, there is hardly any shelf at all. The waters plunge almost straight down from the continents to a great depth. In other places, the continental shelf stretches out 1500 kilometres (930 miles).

The second step is called the **continental slope**, which goes down about 2500 metres. The third part of the ocean is the

The thicknesses and types of sediment on the bottom of our oceans can also be measured. The ocean floor's sediment is up to 7 km (4.3 mi) deep at its deepest point. This is in the Argentine Basin in the South Atlantic Ocean. We can learn a lot about our Earth by studying sediments. Since 1984, international scientists from America's Ocean Drilling Programme have used sediments to learn about the **climate**.

10

continental rise. This is a slope of thick **sediment**, which is made up of rock, soil, **minerals** and the remains of plants and animals.

The fourth part of the ocean is a very deep area made up of flat plains, with many mountains rising from them. The deepest part of the oceans is the Mariana Trench in the Pacific, plunging 11,022 metres. Most mountains lie in chains forming ridges running nearly 6500 kilometres (4038 miles) along the ocean floor. Deep trenches plunge from the ridges, separating the ocean floor into **plates**. The plates are moving apart slowly all the time.

Volcanic activity under the ocean has been partly blamed for what is called the 'El Nino' effect. This is when normal east–west **trade winds** blow the other way, causing **hurricanes**, high temperatures and flooding.

Mountain ranges under the ocean are formed when hot, runny **magma** oozes up from the Earth's **mantle**, beneath the crust. As it cools, the magma forms new ocean crust, pushing the old aside. This process is called sea-floor spreading. Over millions of years the old crust forms the mountains of the mid-ocean ridges. On this map, you can see the Mid-Atlantic Ridge.

The importance of sea water

Oceans and seas have many **minerals** dissolved in their waters. They have gases such as oxygen in them too. All of these are important for plant and animal life under the waves. But oceans and seas are just as vital for life on land. The waters help to control our **climate**. They give us rain too. As the sun beats down on oceans and seas, the heat **evaporates** moisture from the surface of the water. This is carried in the air as **water vapour**. A lot of the water vapour forms clouds which get blown on to the land. The clouds protect our Earth from the heat of the sun. They also bring precious rain.

You can tell the different depths of ocean waters by their colour. Here, along the Great Barrier Reef off the east coast of Australia, the shallow parts near the shore of the island are a lighter colour than the deeper parts. You can see a reef of **coral** in the picture, with pale sea all around it. Coral is made of the shells of billions of tiny sea creatures.

Much of the rain falls on high ground where it gathers in streams and rivers. These flow down to the sea, carrying more minerals from the rocks over which they flow. This movement of water from the sea to the land is never-ending. It is known as the water cycle, or the hydrological cycle. The waters of the oceans and seas also absorb harmful gases from the air. This stops them from rising into the **atmosphere** and spoiling the layers of gases around the Earth.

The seas can freeze! In winter in the polar regions ice begins to form in round patches called pancakes. Icebergs like the one in the picture break away from huge ice sheets, ice caps and **glaciers**. You can only see the tip of an iceberg. Most of it lies under the ocean.

What is sea water like?

There is a lot of salt in our main oceans, and even more in some of our seas, like the Red Sea. There are other chemicals too, such as magnesium, calcium and potassium. The temperature of the water also varies, from 30°Celsius in tropical areas to -1.4°C in the **polar regions**. This is just above the freezing temperature of sea water. But however warm or cold it is, the water's temperature does not change much throughout the day.

The mighty Pacific

The Pacific is the oldest, largest and deepest of our oceans. Some of its rocks were formed at least 200 million years ago. The ocean contains over half of the Earth's free-moving water. The Pacific also holds the record for the deepest point in the world, which is about 11 kilometres (6.8 miles) below the surface. It lies in the Mariana Trench, east of the Philippine Islands.

The Pacific Ocean's northern boundary is the Bering **Strait**. Its southern boundary is the Antarctic. To the east, the ocean laps the west coasts of North and South America. To the west the ocean meets Asia, the islands of Malaysia and Indonesia and the island **continent** of Australia.

This is what happens when a huge wave called a tsunami shatters the shore. Tsunamis are mighty walls of water that swell up when undersea earthquakes rumble along the ocean floor. An earthquake on one side of the Pacific can cause a tsunami on the other side.

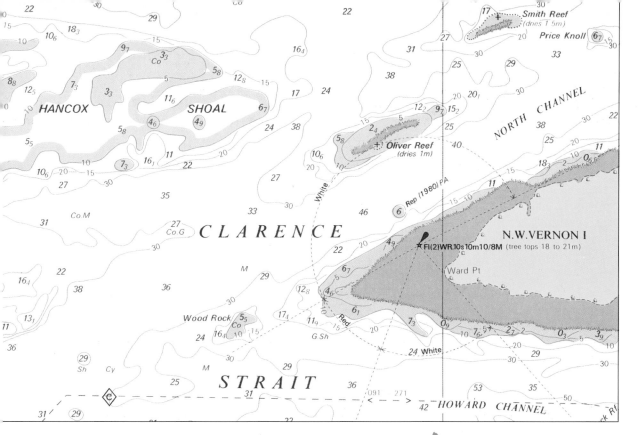

The **continental shelf** (see page 10) is narrow along the coasts of North and South America, but wide near Asia and Australia. The ocean floor is split by wide trenches and studded with chains of mountains.

There are over 30,000 islands rising from the floor of the Pacific Ocean. The South Pacific area has islands made of **coral**. But the western Pacific has a long arc of volcanic islands. This part of the Pacific is called the 'Ring of Fire' because of all its volcanoes and earthquakes.

Coral reefs such as the Great Barrier Reef in Australia are full of sea life, which attracts tourists. In South America, the **currents** have created a good environment for fish. Anchovies are one of the biggest catches. Seabirds also eat the anchovies, and the birds' droppings, called guano, are collected and sold as **fertilizer**.

Here you can see a map showing part of the Pacific Ocean on the coast of Australia. This type of map is called a chart and is used to help navigate ships. The figures over the sea show the depths of the ocean at low tide. They are marked in metres. We can find out whether the sea floor is sandy, rocky or muddy. The chart also shows **buoys**, lighthouses and tall buildings along the coast.

Plants of the oceans and seas

Plants in the water

Green, leafy plants depend on sunlight to make their food energy. So, in the oceans and seas, these types of plant can only grow where the sun's rays reach down into the water. Most of these plants lie in the **continental shelf** area and in the top part of the **continental slope** of the ocean.

The plants are mainly **algae**, which are the oldest form of life able to use sunlight to feed. Algae range from tiny microscopic plants to fleshy-leaved seaweeds such as the giant kelp, which can grow 60 metres each year. The larger algae often have long, thick stalks which are attached to the bottom of the ocean. Some attach themselves to rocks near the shore using clinging roots. They have a slimy coating to stop them going dry when the tide is out.

The sargassum weed has given its name to the Sargasso Sea in the Atlantic Ocean. The weed is really a greenish-brown algae that floats freely in the sea and makes the water look green. The Red Sea is full of a red algae, which can range from pink to orange and a red-brown. There are blue-green and brown algae too.

The smaller plant algae floating in the ocean are also known as **phytoplankton**. They are eaten by tiny creatures called zooplankton, which in turn are eaten by fish. So without algae there would be very little animal life in the oceans and seas. In 1970 scientists found tiny plant **bacteria** growing on rock deep down near the hot volcanic vents in the Pacific Ocean. The bacteria do not need sunlight to feed and give them energy. Instead, they use a chemical called hydrogen sulphide.

Plants near the shore

Plants near the shore have to cope with strong winds and salty air. Some survive in salt marshes where a river meets the sea. Plants that are able to grow in salty conditions are known as halophytes. Many of them get rid of the salt through thousands of tiny holes in their leaves.

The sea buckthorn bush has lots of tightly packed branches and short, thin leaves that do not break in the wind. Its small green flowers have tough brown scales. The flowers and the bright orange berries are clustered closely together against the branches for protection. The bush can grow up to 10 m tall and is found in many parts of Europe.

Sea creatures

The oceans and seas provide a huge range of **habitats** for creatures. These range from the dark, cold depths of the oceans to muddy shores. Like sea plants, most sea creatures live in the sunlit area of the **continental shelf** and the upper part of the **continental slope.**

Full of fish

Some fish are bony and have a mouth with jaws. Others have a skeleton made of a very soft, bendy material called cartilage and a more flap-like mouth. Fish breathe through gills, which absorb oxygen from the sea water into the bloodstream. Fins help fish to move through the water and an air bladder inside the body stops the fish from rolling from side to side. Many fish have a protective layer of **scales**. These are covered with a very thin skin that produces an **antiseptic** slime.

Unlike most molluscs, the octopus has no outer shell. It can grow up to 3 m across and has eight arms. Each arm has two rows of suckers which help the octopus to hold its prey. The octopus has a beak-like mouth with poison sacks near it to stun its catch. The octopus's grey-brown, spotty body can change colour to match its background. This **camouflage** hides it from its enemies. Different species of octopus are found in most oceans and seas.

18

Some fish have adapted to surviving in the deepest, darkest parts of the ocean. These fish create their own light in their bodies. They do this by producing a chemical called phosphorus.

Many molluscs

Molluscs are a huge family of creatures, with about 80,000 different **species**. Salt-water molluscs range from oysters, which lie still on the ocean floor, to the busy squid. They all have soft bodies but most are protected by a hard shell. The space under the shell holds the gills, which are used for breathing and sometimes for collecting food. Small molluscs, like cockles, bury themselves in the mudflats or sand near the shore. They put out a jelly-like feeler called a siphon to suck in water and food.

The porpoise lives in the North Atlantic and Pacific Oceans and prefers the shelter of inlets to the open sea. It is the smallest member of the whale family, which has about 90 different species. Whales are **mammals**, so they breathe using lungs, but they never go on land. The small porpoise has teeth and eats fish. But the huge baleen whale has no teeth and only sucks **plankton** through tiny notches in the mouth bones.

Living with the ocean

The riches of the sea

Fishing communities built some of the world's oldest settlements by the oceans and seas. Many of these lay by deep, sheltered bays, which were good stopping places for trading ships. Large ports developed here along the oceans' busy trade routes.

Venice has been a trading port for 900 years. It lies on the mouths of the rivers Po and Piave, which open out into the Adriatic Sea. The port is protected from the ocean by long barriers of sand called sandbars. Venice was the biggest trading centre with Far Eastern countries such as China. It is now also a modern industrial city, although much industry is concentrated on the mainland around Mestre and Marghera.

Fishing communities provide the world with an important source of food. Some fishermen still catch fish using traditional boats, such as canoes or wooden sailing vessels. But modern methods allow fish to be caught on a very large scale.

Coastal people also farm or collect seaweed. Seaweeds can be used as **fertilizer** and for their vitamins and **minerals**. The jelly-like substance in brown **algae** is used to make colours in paints smooth and to give ice cream its creaminess!

Ice cream, sea, sun and sand have attracted millions of people to visit the coast as tourists. They have changed small fishing villages into huge resorts. People have flocked to the coast to find work in hotels and restaurants. For some countries, coastal tourism is their most important industry.

Marine minerals

Coastal communities also find work taking minerals from the sea. Salt, magnesium and bromine are **evaporated** from the water. Seventeen per cent of the world's petroleum oil comes from the ocean floor. Sand, gravel and oyster shells are dug up from the seabed and used in building. Diamonds are also found in some underwater gravel beds.

◆ This is an oil platform off the coast of Norway in northern Europe. Oil has brought many people to live by the sea. It has changed small coastal communities into big industrial towns. The Norwegians have used the money from oil to set up technological industries on the coast. They know that the oil will not last forever.

A way of life – the Labrador Inuit of Canada

The Labrador Inuit people have lived on the east coast of Canada for centuries. The different Inuit peoples came to North America from the frozen lands of Siberia in Russia over 4000 years ago. They have adapted their way of life to harsh, snow-covered coastlands in winter, and mild, sunny summers. Nowadays, of course, not all Inuit lead traditional lives. Many live and work in North America's towns and cities.

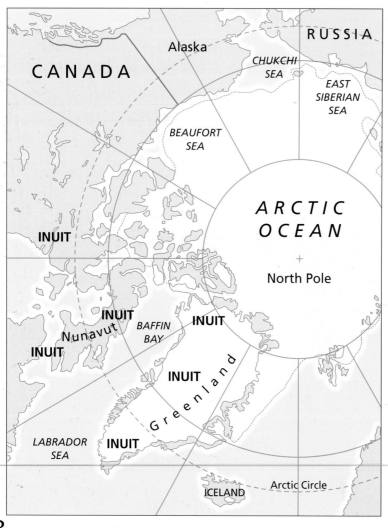

Different groups of Inuit are marked on the map. Every three years Inuit leaders all meet at the Inuit Circumpolar Conference. They discuss issues affecting the region, such as how to preserve the environment. This is especially important for the Arctic Inuit. The Arctic ice is melting because of **global warming** (see pages 24–25). This makes winter hunting difficult. In April 1999 the Inuit of Baffin Island were given their own parliament by the Canadian government. The area has been renamed Nunavut.

Making a living – building a home

Traditional Labrador Inuit use the plants and creatures of the coast and sea for their everyday needs. Seals are a good source of food, especially in winter. The Inuit also eat fish, whales and walruses, which are mostly caught in the summer on the open sea. The meat is freshly cooked, or preserved by drying or freezing.

Seal skins are made into warm, waterproof clothes. Winter trousers and boots are made with double layers of skin and fur. The coat is a parka, which is a double-layer pullover with a hood. The skins are also made into **harpoon** lines, and they cover tent frames and boats as well.

Summer homes use walrus or seal skins over a whalebone or driftwood frame. The walls of winter houses are made of stones packed with moss and earth. Seal fat is burned to make light and heat. Over 300 years ago, Inuit sold carved bone and ivory canes and other goods to European traders. Now they have developed a craft industry using carved soapstone instead.

These days, many people travel on snowmobiles in winter and powerboats in summer. But dog sleds and canoes are still used in some places. Kayaks are lightweight hunting canoes that can go out to sea. They are made of a wooden frame covered with seal skin. The Uliak is a bigger, open boat – about 9 m long and 2.5 m wide. It is used for fishing trips and transporting goods.

Sea changes

Natural changes?

Some parts of the world's oceans and seas nearly always have stormy waters, such as Cape Horn on the tip of South America. Others have long periods of calm, such as the doldrums, which lie in parts of the **Tropics**. But mostly, the ocean waves change all the time. They change with the tides, the weather and the seasons of the year.

In recent years some seas have experienced unusually violent seasonal storms. Severe **hurricanes** have battered islands and coastlines. Many scientists believe that they are caused by too much sun on our oceans. This **evaporates** more water, which brings heavier rain and stronger winds.

The amount of fish in our oceans is falling. We catch at least 60 million tonnes of fish every year. On this fish factory ship, thousands of fish are caught, processed and frozen on board. Sound waves under the water are used to find huge shoals of fish. This technique is known as **sonar**. Lights and electrical pulses attract hordes of fish to the massive nets. Unwanted small fish are often caught and die before they are thrown back into the sea.

Some scientists think that the **climate** is changing because of flares being thrown out by the Sun. Others think the sun is stronger because the Earth's protective layer of ozone gases has got thinner.

This could be caused by harmful CFC gases rising from the Earth into the **atmosphere. Global warming** is thought to be caused by burning too many fossil fuels, such as the petrol we use in our cars and fumes from factories and power stations. Harmful gases are pumped into the atmosphere. Most of these get absorbed by the oceans, seas and green plants. But there is now too much **pollution** for the seas to cope with.

This looks like chemical pollution on our coastline, but it is perfectly natural. During storms, a jelly-like substance in **algae** gets whipped up into a foam. The foam gets washed ashore, where it decays and smells really bad!

What's changing in the water?

Rivers and streams fill our oceans with a continual supply of water and salts. Now they also carry pollutants. Some are chemical **fertilizers** that have soaked into the rivers from farmland. Others are **pesticides**. Raw **sewage** and waste from power stations and factories are sometimes pumped straight into the sea. It is also thought that the stronger sun is destroying **plankton** in the water. Both these things have reduced the numbers of the oceans' plants and animals.

Looking to the future

The power of the sea

The future of the world's oceans and seas depends on us. We can burn fewer fossil fuels, or convert waste gases into cleaner gases. We can also use less energy – or maybe different kinds. The oceans and seas can actually help us to do this. Wave and tide power close to the shore can turn electricity **turbines**. But one of the most recent ideas is to use the ocean's thermal energy. This energy comes from the sun's heat which has been absorbed by the water. It comes from ocean **currents** too. The solar heat can be changed into electric energy in a process known as Ocean Thermal Energy Conversion (OTEC).

This oil spillage came from a huge oil tanker, the Sea Empress, which broke up near Milford Haven in Wales. The oil killed thousands of birds, fish and shellfish. It covered the sands and spoiled an area known for its natural beauty and wildlife. We can stop these kinds of accidents by using different forms of transport for carrying oil.

Fish forever

The number of fish in the seas is falling. More than 90 per cent of the fish caught in the world come from oceans and seas. Many small fish are destroyed before they can become adults and produce more fish. We can stop catching small fish by using nets with larger meshes. We can also develop more fish farms along the coasts, so that fewer fish from the sea are caught.

But there are some things that we cannot help. In the 1970s, the number of anchovies off the coast of Peru began to fall dramatically. Fishermen were blamed for catching too many. But it is now known that this was only partly true. Many shoals of anchovies were moving away from the cold Peruvian Current to find warmer waters.

We can also help ocean life by cleaning the waste from our factories and power stations before it flows into the sea. We can increase the fresh water reaching the sea by building fewer **dams** along the rivers. These are stopping water, and the **minerals** in it, from reaching the oceans.

Some scientists believe that **global warming** is melting ice in the **polar region**. The meltwater raises the level of sea water. This extra water, and increasingly violent storms, are **eroding** our coastlines very quickly. In some parts of the world, towns and villages are falling over the cliffs and into the sea. We need to build coastal defences like these ones in Norfolk, UK, to protect our coastlines.

Ocean and sea facts

Top twelve seas

These are the top twelve seas in order of size. The column on the right shows which of the seas is the deepest. The table also shows which ocean these seas are connected to.

Sea	Location	Area (sq km)	Average depth (m)
South China Sea	Pacific	2,974,600	1464
Caribbean Sea	Atlantic	2,766,000	2575
Mediterranean Sea	Atlantic	2,516,000	1501
Bering Sea	Pacific	2,268,000	1491
Gulf of Mexico	Atlantic	1,543,000	1615
Sea of Okhotsk	Pacific	1,528,000	973
East China Sea	Pacific	1,249,000	189
Hudson Bay	Atlantic	1,232.000	93
Sea of Japan	Pacific	1,008,000	1667
North Sea	Atlantic	575,000	94
Black Sea	Atlantic	462,000	1191
Red Sea	Indian	438,000	538

Plunging the depths

The deepest ocean generally is the Pacific, with an average depth of 4030 m. That's over 4 km! It is closely followed by the Indian Ocean, which has an average depth of 3900 m.

Tsunamis are huge waves caused by undersea earthquakes and volcanic eruptions. One of the most destructive tsunamis happened in 1755 in the Atlantic Ocean. It hit the city port of Lisbon, in Portugal.

The highest wave ever recorded was 34 m high. Here, a surfer is enjoying riding a breaker on the coast of Hawaii, which has some of the biggest waves in the world.

Glossary

abrasion erosion caused by moving stones carried by wind or water

algae simple form of plant life, ranging from a single cell to a huge seaweed

antiseptic substance that kills germs

atmosphere layers of gases that surround the Earth

bacteria tiny, one-celled organisms, some of which can cause disease

basin sea or ocean floor, which slopes downwards like the inside of a washbasin

buoy an anchored float used to mark a channel or obstruction in the water

camouflage way of colouring or covering something so that it is difficult to see against the things around it

climate rainfall, temperature and winds that normally affect a large area

continent the world's largest land masses. Continents are usually divided into many countries.

continental shelf relatively shallow area that slopes from the coast of a continent into the ocean

continental slope the second step under the ocean that runs from the edge of the continental shelf further into the ocean, to a depth of about 2500 metres

coral hard rock made of the shells of tiny dead sea creatures, cemented together with limestone made by the creatures themselves

current strong surge of water that flows constantly in one direction in an ocean

dam wall that is built across a river valley to hold back water, creating an artificial lake

debris eroded material such as rocks, pebbles and sand

erosion wearing away of rock or soil by wind, water, ice or acid

evaporate turn from solid or liquid into vapour, such as when water becomes water vapour

fertilizer substance added to soil to make plants grow better

glacier thick mass of ice formed from compressed snow. Glaciers flow downhill.

global warming gradual increase in temperature which affects the whole Earth. It causes a change in climates.

gravity force that causes objects to fall towards the Earth. We are all attracted to the Earth by gravity.

habitat place where a plant or animal usually grows or lives

harpoon a spear attached to a line, which is often shot from a gun. Harpoons are used to catch whales and fish.

headland cliff sticking out into the sea at the end of a bay

hurricane wind that blows faster than 120 kilometres (75 miles) per hour. Hurricanes can uproot trees and damage buildings.

magma hot, melted rock beneath the hard crust of the Earth

mammal animal that feeds its young with its own milk

mantle layer of hot, molten rock on which the Earth's crust sits

mineral substance that is formed naturally in rock or earth, such as oil or salt

North Pole northernmost tip of the world, in the Arctic region

pesticide chemical that is used to kill plant-eating insects

phytoplankton tiny algae

plankton tiny water creatures that can be either plants or animals

plate area of the Earth's crust separated from other plates by deep cracks. Earthquakes, volcanic activity and the forming of mountains take place at the junctions between these plates.

polar region area around the North and South Poles

pollute make air, water or land dirty or impure

scales small flakes of a hard material rather like finger nails, which cover the skin of many types of fish

sediment fine soil and gravel that is carried in water

sewage human waste material

sonar sound waves which hit an object and then echo back – used to measure distances and depths

South Pole the southernmost tip of the world, in Antarctica

species one of the groups used for classifying animals. The members of each species are very similar.

strait narrow strip of water connecting a sea to an ocean

trade winds winds that blow steadily towards the Equator but which get pulled westward by the rotation of the Earth

Tropics the region between the Tropic of Cancer and the Tropic of Capricorn. These are two imaginary lines drawn around the Earth, above and below the Equator.

turbine revolving motor which is pushed around by water or steam and can produce electricity

water vapour water that has been heated so much that it forms a gas which is held in the air – drops of water form again when the vapour is cooled. There is always water vapour present in the air.

wave-cut platform long, flat step of eroded rock, pebbles and sand that leads to a coastal cliff

Index